Cheers fo

"This is a wine book that everybody should have read before they became intimidated by verbose wine critics. It's basic, practical knowledge written in a logical, straightforward manner for ease of understanding. *Through the Grapevine* should be included in every college graduate's backpack and be in every young professional's briefcase. Give this book to all your friends, share it at your next dinner party, and put *Through the Grapevine* in every Christmas stocking in the house. Cheers, Candace Frasher!"

> Allen Shoup, Founder/Chairman
> Long Shadows Distinguished Wineries and Vineyards

"*Through the Grapevine* is as satisfying as a glass of Beaujolais: refreshing, effortless and pleasurable. Author Candace Frasher breezes easily through the major grape varieties and the wines they make with wit and humor as she reveals the simple essence of good wine and the foods they complement. The clever text and illustrations make you want to turn to the next page as well as reach for another sip."

> Bob Betz, MW, Winemaker/Owner
> Betz Family Winery, Woodinville, WA

"*Through the Grapevine* captures Frasher's down-to-earth and easy to understand teaching style and makes the complex world of wine simple. Read *Through the Grapevine* and never be intimidated by a wine list again."

> Carol Bonino, Director of Foundation Relations
> Gonzaga University, Spokane, WA

Cheers ...

"This publication offers an artistic introduction to some of the most commonly grown wine grapes and their flavors. Newcomers to wine should find *Through the Grapevine* very helpful."

Chas W. Nagel, Professor Emeritus
Washington State University, Pullman, WA

"Playfully thumbing its nose at snobbery, *Through the Grapevine* delivers a usable amount of information with an inventive sense of fun. Candace Frasher instills the reader with the confidence that it is not necessary to know it all. And anyone who can de-mystify the wine choosing process is high in my book."

Don Hamilton, Photographer/Cinematographer
Hamilton Photography & Film Co., Spokane, WA

"Because I know so little about wine, *Through the Grapevine* was a real 'find' for me. Frasher has captured the character of each grape and the wines made from them in a fanciful, yet factual way, giving me what I need to make good selections. Put *Through the Grapevine* on your shopping list for a great hostess gift, birthday remembrance, or thank you present and share this fabulous 'find.'"

Jan Sanders, long time Library Director for
Public Libraries across the land, currently at Pasadena, CA

"Wine snobs, beware! Here is a disarming, unpretentious introduction to wine varietals that captures the spirit of the wine-tasting experience. This lighthearted and elegant presentation makes wine tasting accessible to everyone who's interested in enjoying their food and drink more fully."

Janelle McCabe, Wine Bar Manager, Coeur d'Alene Cellars

"This precious little gem is packed with information that enlightens the lay person's journey into the discovery of the rich world of wine grapes. Frasher combines the muse of creativity with the unique gift of speaking directly, always with charm and wit. Upon finishing this book, I am sure the reader will savor getting to know her with that same warm glow of satisfaction a shared bottle of good wine brings. (Then imagine my warm glow knowing her personally!)"

Ken Spiering, Northwest Artist

"All levels of wine enthusiasts, from the beginner to the experienced connoisseur, will appreciate this charming book. Pick up a copy and enjoy reading Candace Frasher's delightful and insightful descriptions of what makes our business possible … *grapes*!"

Mike & Ellena Conway, Winemaker/Owners
Latah Creek Wine Cellars, Spokane, WA

Cheers ...

"Candace Ann Frasher is nothing short of a savvy wine pioneer with a knack for drawing out the uninitiated wine taster in us all. No stuffy, high 'n mighty wine descriptions here. *Through the Grapevine* offers simplicity as a guiding light. Tasters will delight in playful illustrations filled with fun that linger long after the wine is gone. This book is a gem you won't want to miss."

Peter Rasmussen, Environmental Designer, Costa Mesa, CA

"Written in common man's language and illustrated in watercolor plums, greens, khakis, and ruby reds, *Through the Grapevine* by Candace Ann Frasher is a delight to be held and behold. It proves one need not sweep from aristocracy to be guided into the often-unknown world of wine grapes. *Through the Grapevine* is a must read and keepsake for travelers and life-long learners."

Sean T. Taeschner, M.Ed., **author of**
Finding Gold in Washington State: 2005-6 Edition

"A delightful beginner's book for all up and coming wine lovers who need guidance in a pure and simple, uncluttered manner. Learning the basic grapes together with complementary foods is an excellent start on the journey of understanding the enjoyment of wine. Candace Frasher is a most engaging and entertaining writer."

Shirley Alpert, Director of Communications
House of Burgundy, New York

THROUGH THE GRAPEVINE

An Illustrated Guide to Wine Grapes

By Candace Ann Frasher

Copyright © 2006, 2007 by Candace Ann Frasher. Printed in Canada.

All rights reserved. No part of this book may be reproduced or transmitted in any form or by any means, electronic or mechanical, including photocopying, recording, or by any information storage and retrieval system without written permission from the author, except for the inclusion of brief quotations in a review.

Design: Cherylann Collins
Illustrations: Selina Shehan

For more information on this book and basic wine workshops, please visit www.WineABCs.com.

Library of Congress Cataloging-in-Publication Data
Frasher, Candace Ann, 1948-
     Through the grapevine: an illustrated guide to wine grapes/ Candace Ann Frasher.
        p. cm.
     Includes bibliographical references.
     ISBN-13: 978-0-922993-64-2 (pbk. : alk. paper)
     ISBN-10: 0-922993-64-5 (pbk. : alk. paper)
      1. Grapes--Varieties. 2. Grapes Varieties--Pictorial works. 3. Viticulture. 4. Wine and wine making. I. Title.
SB398.28.F73 2006
634.8'3—dc22   200602365

For David L. Mielke
Visionary, Custodian, Groundskeeper

CONTENTS

Acknowledgments 10
Dedication 11
Author's Note 12
Graphic Designer's Note, Illustrator's Note 13
Foreward 14

Introduction 15
Ground Zero 16

Grapes that Make White Wine 18
    Chardonnay 20
    Chenin Blanc 22
    Gewürztraminer 24
    Muscat Canelli 26
    Pinot Gris 28
    Sauvignon Blanc 30
    Sémillon 32
    Viognier 34
    White Riesling 36

CONTENTS

Grapes that Make Red Wine 38
    Barbera 40
    Cabernet Franc 42
    Cabernet Sauvignon 44
    Gamay 46
    Lemberger 48
    Merlot 50
    Pinot Noir 52
    Sangiovese 54
    Syrah 56
    Zinfandel 58

Wine Notes 60
About the Author, Graphic Designer, Illustrator 66

ACKNOWLEDGMENTS

Cherylann Collins and Selina Shehan, you are incredible women of enormous talent. Thank you for your splendid design and artistry. Janelle MaCabe, master of the English language and passionate wine devotee, thank you for your editing expertise, for clapping where I wrote it right, and for polishing it up where I wrote it rough. Jan Sanders, Jaclin Smith, Marjorie Bender, Gail Goeller, Valerie Patchen, and Marjorie Hinds, bless you for supporting my creative bent. We artists need friends like you. Finally, my heartfelt gratitude to Geneva Morrow Goldizen Frasher, aka mom. You knew I'd never be a fine cook, but you knew I'd be just fine somehow.

DEDICATION

This book is dedicated to the late Paul Gillette, publisher of *The Wine Investor*, who welcomed me into his office in Los Angeles when I was still green as grass in the wine business. I represented a brand new Washington winery and literally had no idea what I was supposed to say or do once I got in the door. Paul graciously accepted the wine samples I brought for review and immediately put me at ease with his wit and charm. When I left Paul's office, I perceived the wine industry in a totally new light. It was going to be exciting, and it was going to be fun. And it has been.

It is in this spirit that I also wish to thank everyone who might have felt intimidated and came to one of my wine workshops anyway. You've made my wine journey immeasurably gratifying by allowing me to share what I've learned. I believe that wine is worth getting to know, that it heightens the pleasure of dining and creates an atmosphere where friends and family linger and laugh and where business associates strategize and make good decisions. I hope to see many of you again. It would be fun to catch up … over a glass of wine.

AUTHOR'S NOTE

In 1989, I began teaching basic wine classes for the Institute of Extended Learning in Spokane, Washington. In an effort to get the information across, I used a variety of supplemental materials because I couldn't find just the right book to recommend to my students. Not to say that there weren't numerous excellent wine books on the market – there were – but the majority of them read like dictionaries. Not much fun. So, in 1990, I decided to write the book that I would want if I were a student in my wine class.

Without an illustrator, graphic designer, and the capital to finance such an undertaking, the copy I wrote for my book lived and almost died in my computer. Before taking its last breath, however, it occurred to me that perhaps university students might like the opportunity to use their imagination, creativity, and talent to work on a "real world" project. How about a wine book?

I ran the idea by my friend, Tom Askman, Eastern Washington University's Art Department Chair. He suggested I contact Mindy Breen, Assistant Professor in the Graphic Communications Department. I rang Mindy and pitched my idea to her. She agreed to meet with me and asked several other professors to join us. In our meeting, I explained what I was trying to accomplish, how this book would differ from everything else on the market, and proposed that students do the design and illustration. After some lively discussion, I got their nod of approval.

Several days later, Cherylann Collins, a student of graphic design, and one of the most talented, charming, focused, and committed females I have ever met, called. Cherylann soon introduced me to classmate Selina Shehan, a gifted illustrator. Words about wine were soon being expressed by images that flowed down Selina's fingertips and onto paper. And thus, the project began.

DESIGNER'S NOTE

Candace, Selina, and I brainstormed the conceptual development of the various characters. After Selina brought them to life on the page, the design of this book started. It was my job to frame the paintings. I wanted these characters to draw the reader in and reflect the passion Candace has for making wine approachable.

The pages are composed in such a way to allow a slight pause as you go from the illustrations to the text. The font used for this book, Adobe Garamond, was chosen because it is classic and timeless. A more ornamental font would have drawn too much attention and focus away from the paintings, which are key. The colors of the borders can be found in the paintings and were selected to portray the personality and intensity of the grapes.

ILLUSTRATOR'S NOTE

The art for this book has gone through many progressions ranging from rough sketches to computer illustration, back to sketches, and finally becoming a style somewhere in between sketch and watercolor. The goal was to capture the fluidity and essence of the wines and their character through quickly drawn pencil lines and layer beauty and richness through the use of watercolor paints. Ultimately, I picked some key moments and people in my life that inspired me and tried to convey that sense of awe on the blank page.

Special thanks to Candace, Cherylann, and to my mother, Linda, whose energy and spirit embodies everything that a great wine is really about.

FOREWARD

Through the Grapevine is exactly the kind of book that so many of us looked for when we first began to discover wine, but it didn't exist until now. The timing of its publication couldn't be better. In 2007, the United States will become the largest consumer of wine in the world and to that milestone there must be scores of people discovering the joys of wine for the very first time. These new wine drinkers are learning what Europeans have known for centuries … that wine is consumable art with the ability to teach us about food, geography, and tradition while simultaneously enhancing our quality of life on a daily basis. There is no one better to write this book than Candace Ann Frasher who put wine and art together in *Through the Grapevine* and whose passion to learn and share is only eclipsed by her contagious zest for living a full, generous, and compassionate life. Cheers!

Steve Burns, Principal
O'donnell lane LLC
Glen Ellen, CA

INTRODUCTION

For the Love of Grapes

I love grapes. They grow in clusters. These perfectly elliptical spheres are remarkably sensuous with skins touching and juices plumping up their insides.

Grapes take delight in the position Mother Nature has given them. From early morning until late afternoon, they bask in the sun. Some varieties steadfast and true, others untamed and persnickety, yet all lay naked under blue skies soaking up the sun's rays. In the cool of the evening, these little round wonders feel their racy acids. Come morning, it's time to sun themselves again. Day by day, they feel the magic of the sun's light and take delight in becoming plump and sweet. At the end of summer when the kiss of fall turns the earth vibrant colors and hues of red and gold appear, sumptuous clusters boast joyfully as crisp acidity and sugar sweetness come together in perfect balance.

Let the harvest begin. Luscious grapes that make the world's finest wines have once again reached their full potential as ripe, juicy orbs.

GROUND ZERO
WINE IS MADE FROM GRAPES

Wine, a traditional beverage of numerous cultures around the world, is often celebrated as a gift of nature to humankind. It is made from many different grape varieties, all having distinct characteristics that result in wines with unique textures, aromas, and flavors.

In the United States, most wines are named after the particular grape variety that the wine was made from. For example, Chardonnay, Cabernet Sauvignon, and White Riesling are the names of wine grapes. The wines made from these grapes are named the same.

Because so many wine grapes originated in European countries, the French, Germans, and Italians, etc., had the privilege of naming them. This foreign nomenclature has made it tough for Americans to read wine lists, pronounce the names correctly, and order a bottle of wine. Therefore, if you stumble and mumble at first, be patient. Being tongue-tied is simply an inescapable part of the learning curve.

WINE AND FOOD PAIRINGS

I advise experimenting with gusto when matching wine and food. No matter what you've heard, there are no hard and fast rules. The whole idea is to use wine as you would a condiment to enhance and complement your meal. The trick is balancing the flavors so that neither the wine nor the food overwhelms the other. It's really very simple. Take a bite of food. Now, take a sip of wine. If it tastes good – it is! You are the judge.

To get you started, I've made some suggestions that are usually pretty tasty; however, if you feel like experimenting, please do. Exquisite matches are endless and your favorites are awaiting you!

GRAPES! GRAPES! AND MORE GRAPES!

Although an astounding number of grape varieties grow on this planet, there are only a handful that can be considered super stars of the wine world. The following pages will show you the major players. One thing is for sure: if you learn these, you'll have a running start.

GRAPES THAT MAKE WHITE WINE

The following grape varieties make some of the world's most sensational white wines. The clusters of most varieties are actually a lovely light green and the color of the wine can be anywhere from a pale straw to brilliant gold. Referring to these beauties and the wines they make by the traditional term "white" doesn't really do them justice, as your eyes will tell you.

Pouring gold; rich, glamorous ... Chardonnay

Chardonnay

(Shar-doe-nay')

Chardonnay laughs in the vineyard, sings in the winery, and dances in the glass. Truly an ancient vine variety, Chardonnay grapes make one of the most sought after dry white wines in the world today. The wine virtually flows in every major wine producing country due to the grape's ability to grow in a variety of climates and the demand for this luscious, elegant wine by its fans. If your family enjoys traditional holiday fare complete with turkey, gravy, and the trimmings, bring on the Chardonnay. It will complement the main course and a wide variety of side dishes such as scalloped corn and winter squash. All together now ... Cheers!

My impressions of this wine: _____

_____

_____

_____

A wine of many styles ... Chenin Blanc

Chenin Blanc

(Sheh'-nin Blahnk)

Chenin Blanc*, a French native, is a magical chameleon of a grape. It produces exquisite sweet wines, nervy dry wines, and fine sparkling wines. On warm summer evenings, serve Chenin Blanc in one or more of its fashionable styles with an assortment of delectable appetizers, patés, and dips.

* The word "blanc" means *white* in French. Unlike the English language, final consonants in French are often silent. The word "bouquet" is a good example. Thus, when in France, do as the French do and say "blahn" instead of "blahnk!"

My impressions of this wine: _____

_____

_____

_____

Flirtatious, enchanting, exotic ... Gewürztraminer

Gewürztraminer

(Gah-verts'-trah-meener)

Even from a distance, Gewürztraminer is an easy grape to recognize since its clusters are neither green nor red but most definitely pink. The wine is easy to recognize as well. Intensely perfumed, it is outcast by some and much adored by others. Like this wine or not, learning to pronounce Gewürztraminer is a nuisance. It's often helpful to remember that the word "gewürz" actually means *spice* in German. When you crave flavor, try Gewürztraminer with Indian or Asian foods as the spiciness of the wine and the spiciness of these cuisines work together perfectly.

Pronunciation tip: Pour a glass of Gewürz. Take a sip; taste the spice; pronounce the name. Take another sip; taste the spice; pronounce the name. Repeat until you have perfect pronunciation, or you think you do!

My impressions of this wine: _____

_____

_____

_____

Luscious all by itself ... Muscat Canelli

Muscat Canelli

(Muss'-cat Kah-nell'-ee)

Known as "Muscato de Canelli" in Italy (roughly translated as "muscat grapes from the village of Canelli"), this wine is powerfully fragrant and tastes like ripe, juicy grapes that were just picked off the vine. It can be a lovely accompaniment to fruit desserts or totally satisfying when sipped and savored alone. Once met, Muscat Canelli can never be forgotten.

PS: Don't let the name confuse you. A Muskrat is a furry, aquatic animal. None has been reported to enjoy this wine.

My impressions of this wine: _____

_____

_____

Exquisite with "fruits of the sea" ... Pinot Gris

Pinot Gris

(Pee'-noh Gree)

Pinot Gris prefers growing in cool climates and is especially well suited to the Oregon coast, Alsace in northeastern France, and northern Italy (where it is called Pinot Grigio). Pinot Gris is definitely at its best as a dry wine to accompany food and is always a good choice with seafood. Don't let this one get away; order "fruits of the sea" and Pinot Gris!

My impressions of this wine: _____

_____

_____

_____

Racy, aggressive ... Sauvignon Blanc

Sauvignon Blanc

(Soh'-vee-nyohn Blahnk)

Sauvignon Blanc grapes were planted in California over 100 years ago. Making wines whose character is closer to flaunting the Salsa than Waltzing Matilda, Sauvignon Blanc* is a wine of our times: white, dry, and full of zest. You don't ponder this one. Its pleasure lies in its insistent, straightforward flavors of new mown grass and tropical fruit. If you're an oyster fan or scallop devotee, this wine and food match is for you.

*This wine is also called Fumé Blanc *(Foo'-may Blahnk)*.

My impressions of this wine:

Making tracks ... Sémillon

Sémillon

(Say'-mee-yohn)

Sémillon, an ardent traveler, has put down roots all over the world. Wines made from Sémillon can be made in a dazzling array of styles; dry and fresh, light and fruity, or sweet and honeyed. Enjoy the dry versions with saucy chicken dishes and the sweet ones with fresh fruit or a rich, creamy blue cheese. So many choices, take your pick!

My impressions of this wine: _____

Grapes that Make White Wine

A peach of a grape, the very fashionable ... Viognier

Viognier

(Vee'-oh-nyay)

Try saying Viognier. It feels good. The wine made from Viognier tastes good, too. It may be a newcomer to many American palates; however, it is making a splash on more and more dining room tables. The attention-getter is Viognier's lush peach perfume and peachy, apricot flavors. Drink it when you're young and start pronouncing it now. By the time you're old, the wine will be a favorite and how you pronounce it won't matter.

Order Viognier with grilled halibut or sea bass and risk becoming a fanatic. There are many less worthy obsessions.

My impressions of this wine: _____

_____

_____

_____

High on charm and fruit ... White Riesling

White Riesling

(Rees'-ling)

White Riesling, or simply called Riesling, is a noble grape of German descent. Experts agree that these grapes make some of the finest white wines in the world. Rieslings are splendid when served as an aperitif or as an accompaniment to fish dishes such as trout or sole. Order them without hesitation and joyfully anticipate the lovely fruit flavors that only Rieslings can offer.

Note: Schloss Johannisberg is Germany's most famous vineyard and estate. Its Riesling wines were so renowned that many wineries in the United States named their Rieslings "Johannisberg Riesling" after Schloss Johannisberg. This practice is becoming a thing of the past and most U.S. wineries are now naming their wines White Riesling or Riesling after the grape. (The German word "schloss" means *castle* and "Johannisberg" means *Hill of St. John*.)

My impressions of this wine: _____

_____

_____

GRAPES THAT MAKE RED WINE

Many people have to acquire a taste for red wine, i.e., you have to learn to like it! It may take awhile, but once you are charmed, be prepared, because there's no turning back. As red wine lovers like to say, "All white wine would be red if it could!"

Personality plus ... Barbera

Barbera

(Bar-bair'-ah)

Not unlike certain personalities, the grape Barbera makes a wine that is opinionated but likeable. In the glass, this heady red tends to be big, rich, and robust while its structure is round and smooth. Sound like anyone you know? Pour Barbera and serve up a plate of your favorite spaghetti with an herb or meat sauce for a combination that is sure to please.

My impressions of this wine: _____

_____

_____

_____

A star is born ... Cabernet Franc

Cabernet Franc

(Ca'-bair-nay Frahnk)

Because of Cabernet Franc's unique characteristics, this grape has long been used to make a wine to blend with other red wines, particularly Cabernet Sauvignon and Merlot. It is no longer simply a stagehand, however. Made on its own, Cabernet Franc is now the star of the show. Show it off with barbecued ribs and hands will clap.

My impressions of this wine: _____

_____

_____

Intense, dynamic, and compelling ... Cabernet Sauvignon

Cabernet Sauvignon

(Ca'-bair-nay Soh'-veen-yohn)

The Cabernet Sauvignon grape makes the most lordly, powerful, red wines in the world. Its richness, complexity, and depth can be totally seductive. This is a wine that beginning wine drinkers may need to flirt with on occasion before they are smitten, but don't be shy. Cabernet Sauvignon, a classic with beef, is worth the effort to get to know.

My impressions of this wine: _____

_____

_____

_____

Young, unpretentious ... Gamay

Gamay

(Ga-may')

Gamay is the splendid, irreverent red grape grown in the Beaujolais *(Boe-zhoe-lay')* region of France. It makes a very friendly, pleasant red wine that is perfect for taking on spontaneous afternoon picnics in the park. Don't be surprised when drinking Gamay Beaujolais* if visions of bucolic Frenchmen wearing black berets and red striped T-shirts appear.

*The wine made from the Gamay grape in Beaujolais is called Gamay Beaujolais!

My impressions of this wine: _____

_____

_____

_____

Grapes that Make Red Wine

Simply delicious ... Lemberger

Lemberger

(Lem'-burger)

In this case, Lemberger only sounds like cheese! Washington State takes the prize in the United States for successfully growing the little known grape variety, Lemberger. It makes a bright red wine full of yummy, plummy fruit flavors. When you're in the mood to experiment, try Lemberger with barbecued burgers or chicken. Give Limburger, the cheese, a go only if you have a passion for raucous explosive flavored, soft cheese.

My impressions of this wine: _____

_____

_____

Soft, alluring ... Merlot

Merlot

(Mare-low')

Although Merlot's homeland is France, this red beauty found its way to the United States and decided to call this country home as well. Merlot's suppleness and charm often make beginning red wine drinkers passionate devotees. It is soft and round, a red wine that's easy to warm up to. Match it with pork or veal and you'll be reaching for more.

My impressions of this wine: _____

_____

_____

_____

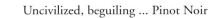

Uncivilized, beguiling ... Pinot Noir

Pinot Noir

(Pee'-noh Nwahr)

Pinot Noir* is the most finicky of the great red grapes. Each year, growers must approach these vines with courage since it is an exasperating, demanding, and infuriating grape to grow. The wine, however, truly tantalizes with its riches, so growers put up with it. We red wine drinkers thank you growers for your persistence in the vineyard and encourage you to dream on.

As an alternative to matching white wine with fish, pair Pinot Noir with salmon or seared halibut. It's a winning combination. Imagine that!

*The word "noir" means *black* in French.

My impressions of this wine: _____

_____

_____

Grapes that Make Red Wine

A full-blooded Italian ... Sangiovese

Sangiovese

(San-geo-vay'-zeh)

Sangiovese is Italy's most widely planted red grape. If you've ever tried Chianti, you've tried Sangiovese. This medium-bodied, fruity red wine is a natural to pair with Italian food. Why not take a quick trip while you're relaxing at home? Put in a CD by Andrea Bocelli, order a pizza, and open a bottle of Sangiovese. Pour a glass and relax in a cozy chair. Now, take a sip, close your eyes, and dream of Italy. A few sips more, the pizza arrives, and you're there!

My impressions of this wine: _____

_____

_____

_____

Grapes that Make Red Wine

Dry, dark, dense ... Syrah

Syrah

(Syr-ah')

Syrah* is a brute of a grape. It makes a beautiful, deeply colored red wine that has startling taste sensations. When you're not in a hurry and want to spoil yourself, try Syrah. You definitely need time to linger over this one. A bottle of Syrah and an oven-roasted rack of lamb make a dynamic duo that will not disappoint.

* The Syrah grape is called "Shiraz" by the friendly folks Down Under. G'day mate!

My impressions of this wine: _____

_____

_____

_____

Thrilling, mysterious, California ... Zinfandel

Zinfandel

(Zin'-fon-dell)

Zinfandel*, once grown almost exclusively in California, is now found in other grape growing regions in the United States. It is a rich, beguiling, and spicy red. Those who fall for the charms of "Zin," fall hard. Marry spicy foods like jambalaya or hard cheeses such as aged cheddar and Gouda with this big red. You'll find these combinations work perfectly together … forever.

*This grape was responsible for the blush wine, "White Zinfandel," phenomenon in the United States.

My impressions of this wine: _____

_____

_____

_____

WINE NOTES

WINE NOTES

WINE NOTES

WINE NOTES

WINE NOTES

WINE NOTES

ABOUT THE AUTHOR
Candace Frasher

Candace Frasher has a B.A. in Zoology from the University of Washington, an M.S. in Communications from Eastern Washington University, and has worked in the wine industry for over 20 years. She began her career in Washington State and sold Washington wine to buyers in the international marketplace. For nine years, Candace taught basic wine classes offered by an extended learning program of a community college. She now holds wine workshops for graduate students and young professionals, conducts wine seminars for corporations, and holds private wine classes for small groups.

ABOUT THE GRAPHIC DESIGNER
Cherylann Collins

Cherylann Collins graduated Summa Cum Laude with a B.A. degree in Visual Communication Design from Eastern Washington University. She keeps busy balancing design interests and her family, a husband and two young daughters. She is also a passionate photographer and web designer.

ABOUT THE ILLUSTRATOR
Selina Shehan

Selina Shehan is an award-winning graphic designer and illustrator. In addition to being the graphic designer for Rings and Things, Selina designs logos, web sites, and illustrates as a freelance artist. She lives in Spokane, Washington, with her two chihuahuas, Spike and Lola.